SAVING THE TIGER

SARAH EASON

CHERITON
CHILDREN'S BOOKS

Published in 2023 by **Cheriton Children's Books**
PO Box 7258, Bridgnorth WV16 9ET, UK

© 2023 Cheriton Children's Books

First Edition

Author: Sarah Eason
Designer: Paul Myerscough
Editor: Jane Brooke
Proofreader: Tracey Kelly
Consultants: David Hawksett, BSc and Richard Spilsbury

Picture credits: Cover: Shutterstock/TaraWilllow Photography. Inside: p1: Shutterstock/Kirsanov Valeriy Vladimirovich; pp4-5: Shutterstock/Vladimir Wrangel; pp6-7: Shutterstock/Eric Isselee; p7b: Shutterstock/Clarst5; p7l: Shutterstock/BearFotos; p7r: Shutterstock/Eric Isselee; p7t: Shutterstock/Eric Isselee; p8b: Shutterstock/Vaclav Sebek; pp10-11: Shutterstock/Jong Kiam Soon; pp12-13: Shutterstock/Jan Stria; p13b: Shutterstock/Mike Bd; p13c: Shutterstock/Sourabh Bharti; p13t: Shutterstock/Matt Gibson; pp14-15: Shutterstock/Kirsanov Valeriy Vladimirovich; pp16-17: Shutterstock/Gudkov Andrey; p17b: Shutterstock/Monkey Business Images; p17t: Shutterstock/Monkey Business Images; pp18-19: Shutterstock/Greenaperture; p19b: Shutterstock/Sarah Cheriton-Jones; pp20-21: Shutterstock/Gregory Zamell; p21b: Shutterstock/Meunierd; p21c: Shutterstock/Thorsten Spoerlein; p21t: Shutterstock/Captain Al; pp22-23: Shutterstock/Archna Singh; p22b: Shutterstock/Morrowlight; p22t: Shutterstock/Morrowlight; pp24-25: Shutterstock/Sunsetman; p25b: Shutterstock/Attila Jandi; pp26-27: Shutterstock/Bhasmang Mehta; p27b: Shutterstock/Slowmotiongli; p27c: Shutterstock/Sarah Cheriton-Jones; p27t: Shutterstock/Anuradha Marwah; pp28-29: Shutterstock/CherylRamalho; p29b: Shutterstock/Prostock Studio; p29c: Shutterstock/Prostock Studio; p29t: Shutterstock/Sambulov Yevgeniy; pp30-31: Shutterstock/Soumyajit Nandy; p31b: Shutterstock/Soumyajit Nandy; pp32-33: Shutterstock/Mukkesh Photography; p32b: Shutterstock/Soumyajit Nandy; p33b: Shutterstock/Spotters Studio; p33c: Shutterstock/Dangdumrong; p33t: Shutterstock/Martin Mecnarowski; pp34-35: Shutterstock/Etapok; p35b: Shutterstock/Christine Glade; p35c: Shutterstock/Christine Glade; p35r: Shutterstock/Claudia Ferrer; pp36-37: Shutterstock/Andywak; p36b: Shutterstock/Igor Anfinogentov; pp38-39: Shutterstock/Maggy Meyer; p38b: Shutterstock/Bildagentur Zoonar GmbH; p39b: Shutterstock/Jo Reason; p39c: Shutterstock/Abeselom Zerit; pp40-41: Shutterstock/Enky03; p40b: Shutterstock/Gelpi; p40t: Shutterstock/Gelpi; p42t: Shutterstock/Sergey Panikratov; p43b: Shutterstock/Mark Gusev; p44t: Shutterstock/Popova Valeriya; p45b: Shutterstock/Fizkes.

Printed in China

Publisher's Note: The information in the Kids on a Mission features in this book are suggestions for actions that children can take to help protect endangered animals, based on extensive research by the author and consultant. The email addresses and the children featured in the photographs are for illustrative purposes only.

Please visit our website,
www.cheritonchildrensbooks.com,
to see more of our high-quality books.

CONTENTS

TIGERS IN DANGER

Tigers are awe-inspiring animals.
These kings of cats are fearsome killers,
which is why people are so scared of them.
But tigers are far more at risk than humans.
In fact, tigers should be scared of us!

HUMANS ARE THE THREAT

One hundred years ago, around 100,000 tigers lived on Earth. Today, only around 3,900 tigers can still be found in the wild. That is because tigers have been hunted in huge numbers by humans. Every year, hundreds of tigers are killed for their fur and bones. Their homes have also been destroyed because of people. They are cutting down the forests and **grasslands** where tigers live to make room for homes, **mines**, and farms.

The International Union for **Conservation** of Nature (IUCN) keeps a record of the world's **species** and how at risk of **extinction** they are. It is called the **Red List.**

There are more than **142,500** species on the Red List.

The tiger is listed as **endangered.**

LOST FOREVER

Some types of tigers have already died out. The Javan, Caspian, and Bali tigers have not been seen for more than 40 years. They are now believed to be extinct. **Conservationists** are warning us that we need to do more to save the remaining tigers on Earth. If we don't, by the year 2050, they, too, may be gone forever.

HELP THE TIGER!

But there is still hope for the tiger, and it is not too late to save it. Around the world, people have heard the tiger's roar for help. And they are making it their mission to help these amazing animals survive. In this book, we'll learn about the tiger and why it is in danger. We'll discover what people on a mission are doing to help tigers and how they have built a career in conservation. We'll find out how kids everywhere can make it their mission to help save the tiger. And we'll learn how you can make a career in conservation your mission. Feeling mission-ready? Then read on!

◄ Tigers have lived on Earth for millions of years. If we lose this incredible animal, we'll lose one of the most magnificent creatures ever to roam our planet.

Joseph Vattakave, tiger **biologist**

CHAPTER ONE

MEET THE TIGER

Tigers are incredible creatures. They are the world's biggest cats and are deadly killing machines. When fully grown, these huge beasts can weigh up to 550 pounds (249 kg). They are more than 100 pounds (45 kg) heavier than an adult lion!

CATS WITH STRIPES

Unlike other big cats, a tiger has stripes. These striking black marks cover the tiger's orange fur. They help the cat hide in the forests and grasslands in which it lives. Every tiger has its own special stripe pattern. That helps scientists and conservationists identify each individual animal.

TIP TO TAIL

From its nose to its tail, an adult tiger measures around nine feet (3 m) long. A tiger's tail alone can measure three feet (1 m) long! Tigers have enormous heads with round ears and a pink nose. A tiger's **sense** of smell is much better than that of a human, and it uses it to track down its **prey**. The legs of a tiger are incredibly muscular. The animal uses them to run and chase prey and then pounce on it. It also uses its powerful legs to climb trees.

FIERCE FIGHTERS

Tigers are fierce hunters and fighters. They **stalk** their prey quietly through the forest. They can run quickly over short distances to catch prey. They use their powerful paws and claws to grasp hold of it. Then, they sink their deadly fangs into the animal's throat or neck to kill it.

TEETH AND CLAWS

Tigers have huge paws that are around six inches (15 cm) wide. At the end of these enormous paws are long and lethal claws. Tigers also have four incredibly long fangs. Two are on the top jaw, and two are on the bottom.

In a Group

Scientists group animals to help them classify, or order, them. Tigers belong to a group of animals called *Panthera*. Other big cats are also members of the group. They are lions, leopards, jaguars, and snow leopards.

WHERE TIGERS LIVE

Tigers once lived in large areas of Asia. They could be found across the continent, from Turkey, India, Cambodia, and Malaysia to Indonesia, China, and eastern Russia. Today, tigers are found in just a small part of the **territory** in which they once lived. Most are found in India and Thailand. A few still live in Siberia and countries in Southeast Asia.

TYPES OF TIGERS

There are nine different types of tigers. These include the following:

- Caspian tigers
- Bali tigers
- Javan tigers
- Bengal tigers
- Indochinese tigers
- Sumatran tigers
- Malayan tigers
- Siberian tigers
- South China tigers

TOO LATE TO SAVE

We know that Caspian, Bali, and Javan tigers are already extinct. Scientists also believe that the South China tiger may be extinct in the wild. This tiger used to roam across several parts of southern China. However, no one has seen it in the wild for more than 10 years. Scientists also believe that tigers are more at risk in some countries than in others. The number of tigers in Myanmar, Cambodia, Laos, and Vietnam is dropping fast. There are now so few tigers in Myanmar and Laos that it may already be too late to save them.

◀ Tigers have different physical features dependent on where in the world they live. For example, a Siberian tiger (shown left) has a very thick coat of fur to help it survive the freezing winters of Russia.

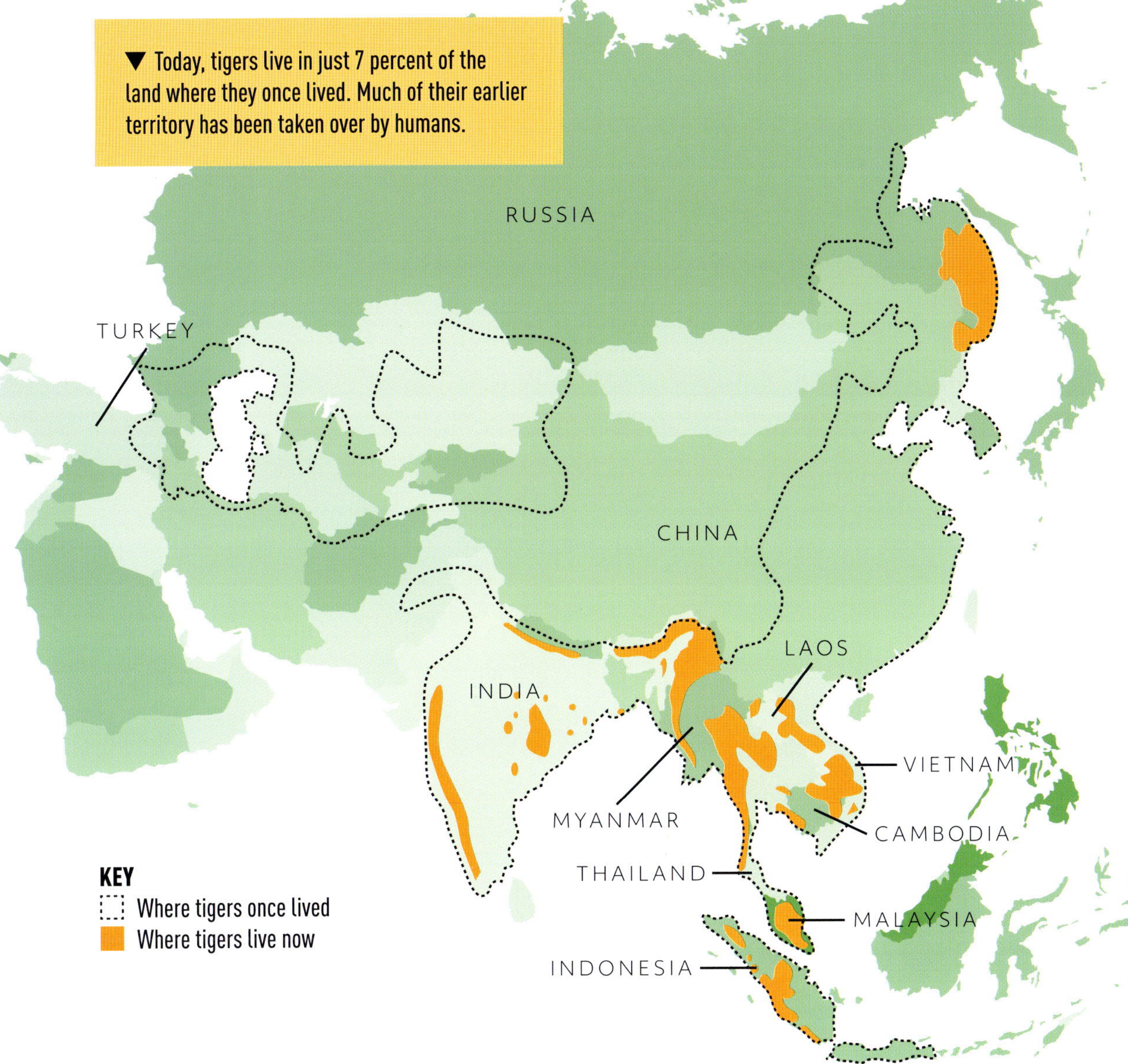

STILL SURVIVING

More than half of the tigers that remain in the wild are Bengal tigers. They live in different types of forests in India. They are also found in the mountains of northern India and the country's **rain forests** toward the south. Tigers like to live near water. Many Bengal tigers are found in the swamps around the mouth of the Ganges River.

9

WHY SHOULD WE HELP THE TIGER?

Tigers are incredibly important to the **environments** in which they live. Tigers are apex **predators**. That means they are at the top of their **food chain**. They hunt other animals in the food chain and have no natural predators.

IN CONTROL

Tigers help control their ecosystem, because they hunt animals within it that would otherwise become too great in number. Animals that tigers hunt are herbivores. These are plant-eating animals. If they become too numerous, they could eat too many of the plants they feed on. These plants would then die out. That would leave no food for herbivores. Then, they would die out, too.

NOT JUST THE TIGER

Many different **organisms** live in the forests and grasslands in which tigers are found. If these places are kept healthy for tigers, they are kept healthy for every other organism that lives there, too. Protecting tigers also helps people. The **habitats** in which tigers live provide people as well as plants and animals with precious food and water.

◀ If the tiger disappears because it loses its home, so too will hundreds of other organisms.

TOP OF THE FOOD CHAIN

Tigers are powerful creatures that are built to hunt. They are one of nature's killing machines. They are efficient and supersmart hunters.

CAT CHASE

If a tiger needs to chase its prey, it can run very fast. Its long, strong tail helps to balance the animal's body as it gives chase.

DEADLY IN THE DARK

Like all cats, tigers can see well in the dark. This makes them deadly nighttime hunters. Their eyes also seem to glow in the dark! A special **adaptation** allows them to do this. A layer in the tiger's eye **reflects** more light than that in most other **mammals**. That makes the eyes appear to glow.

SMART CAT

The tiger is a smart hunter. When it spots its prey, it lies in wait for it. As soon as it is sure it can **ambush** its prey, the tiger leaps out and sinks its teeth into the victim.

ON THE MENU

A tiger's favorite foods are hooved animals, such as deer. However, tigers will eat anything that they can catch and kill. This includes monkeys, turtles, baby elephants, and even crocodiles!

LONG AND LETHAL

Tigers are armed with long, lethal fangs that stab into the flesh of their prey. They also have rows of sharp teeth that slice through flesh and bone.

13

PEOPLE ON A MISSION

Many people around the world are trying to save the tiger. They have made it their mission to help the magnificent animal survive. Many have found careers in conservation as a way to protect this special species. These are some of the ways in which people are working to protect tigers.

PEOPLE IN OFFICES

People in conservation organizations work with governments in countries where tigers live. Together, they create systems that protect the tigers and their habitats. There are also charities in which people work hard to raise money to save tigers. **Communications** and public relations experts help organizations raise awareness about tigers.

PEOPLE ON THE GROUND

Protecting the tiger's habitat is the number one thing that people can do to help it. Wildlife managers and officers are conservationists who work in places in which the tiger lives. They help protect its home and monitor its numbers. Public educators and outreach specialists work with people to help educate them about tigers. They encourage people to protect tigers and make their homes a safer place for them to live in. Wildlife inspectors and **forensic** specialists look for clues about tigers and use science to analyze them.

PEOPLE EVERYWHERE

Ordinary people everywhere, including kids just like you, are raising money to save tigers. They are trying to educate others about the threats tigers face. They are joining organizations that help tigers. They are changing their lifestyles to protect tigers, too. They want a world in which tigers are still around for their children to see.

MAKE IT YOUR MISSION

You can help tigers and other endangered animals by taking action and planning a career in conservation. Here's how:

1. In this book, you'll discover what actions kids on a mission can take to help tigers. Use them to inspire your own actions to rescue tigers.
2. You'll also discover some of the careers people on a mission have in tiger conservation. As you read about each one, consider if that career in conservation might suit you.
3. At the end of this book, you'll find a guide to how to build a career in conservation. Check it out to discover how you can make saving animals your life mission.

▲ This scientist is studying tracks left by a Siberian tiger.

15

CHAPTER TWO

HOMES UNDER THREAT

Tigers have lost a huge area in which they once lived, and they are continuing to lose their homes at an alarming rate. Today, tigers have just a small percentage of the land they once roamed across 100 years ago. What is happening to these magnificent creatures?

TOO MANY PEOPLE

Unfortunately for tigers, there are now more people on Earth than ever before. And they all need somewhere to live and food to eat. In India and Southeast Asia, **populations** are getting bigger and bigger every year. As populations grow, forests and grasslands in which tigers live are cleared to make way for homes and farmland. New roads are built that cut through the tigers' territory. Towns and villages are built on their land.

TAKING OVER TERRITORY

Tiger territory is also being taken over by industry. People are cutting down the trees in forests for **timber**. They are also clearing the land on which tigers live to grow **plantations** of palm oil, rubber, and other crops. These are then sold to people all over the world.

Mining companies are moving into tiger territory, too. People are mining for **fossil fuels** such as coal, metals such as iron, and **minerals** such as talc. The mining companies taking these **resources** use powerful machinery to do so. They bulldoze their way across tiger territory. They cut down trees and grasslands, and dig into the ground. Large areas of tiger land are being destroyed due to these activities.

RESERVED FOR TIGERS

To help protect tigers, conservationists have set up wildlife reserves. These are areas in which tigers should be safe from threats such as farming, mining, and logging. Wildlife reserves are monitored by people working in tiger conservation, who try and keep the tigers there safe.

◀ Rain forests and grasslands are perfect hunting grounds in which tigers can stalk their prey.

When I found out how the tiger's home is being destroyed, I knew I had to help. I asked my teacher if I could do a research project on tigers. Then, I showed it to all my classmates. Now, they are helping by telling their friends about tigers.

"With enough ... habitat, prey, and protection ... tigers can make a comeback."

Stuart Chapman, World Wildlife Fund (WWF)

FARMING

Farming is a major problem for tigers. Farmers clear tiger territory to grow crops such as rice, wheat, and a root vegetable called cassava. The land is also cleared so that farmers can raise livestock on it. When land is cleared for farming, the animals that tigers feed on may disappear. The tigers must then look elsewhere for food.

TIGERS GOING HUNGRY

Most tigers are shy animals and generally avoid people. However, when tigers cannot catch enough prey, they get hungry and desperate. They may attack and eat livestock, such as cattle and sheep. Farmers angered by losing their livestock shoot tigers. Local people are also terrified when tigers enter their villages looking for food. Sometimes, people have been attacked by tigers. They are therefore naturally frightened of the animals. They, too, kill the tigers out of fear to protect themselves and their families.

PEOPLE ON A MISSION: TIGER TEACHERS

Public education and outreach specialists teach people about tigers and their habitats. They work with local people to educate them about tigers and the importance of protecting them.

Specialists work with local communities in tiger habitats to try and stop **conflict** between them and tigers. Specialists educate farmers. They explain that tigers are attacking livestock because they have lost their food supply. Specialists work with farmers to encourage them to move their farmland away from tiger territory. That gives tigers more land on which to roam and hunt, which keeps them from hunting livestock.

Specialists also teach farmers about technology that can protect their livestock. Special lighting, noisemakers, and high-tech fencing can all help keep tigers from attacking farm animals. By using these technologies, farmers can reduce tiger attacks.

Specialists and organizations work with local communities to stop tiger attacks in villages. They employ locals to patrol villages. They train them to react to tigers without killing them. For example, tigers can be kept away from villagers using nets, sticks, and noisemakers. Locals then alert people at wildlife authorities. They can then capture the tigers and safely return them to their habitat.

◀ This farmer is cutting her wheat crop in India. Conflict between farmers and tigers has been a problem in the country for years.

Hungry tigers have sometimes ▶ attacked farmers. This has led to an understandable fear of the animals among local people.

SHARING ITS HOME

The tiger shares its grassland and forest home with other amazing animals. Many of them are endangered, too.

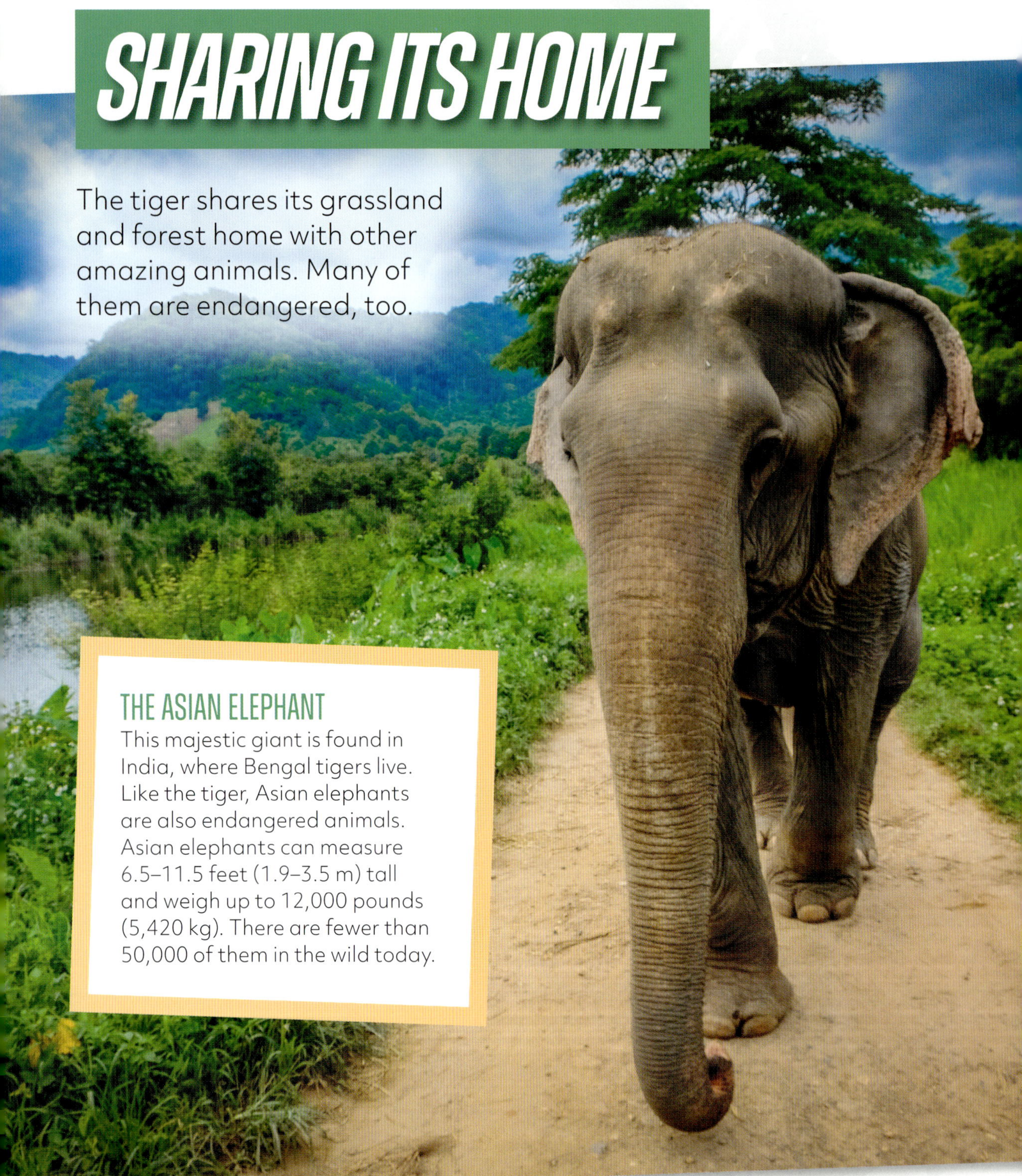

THE ASIAN ELEPHANT

This majestic giant is found in India, where Bengal tigers live. Like the tiger, Asian elephants are also endangered animals. Asian elephants can measure 6.5–11.5 feet (1.9–3.5 m) tall and weigh up to 12,000 pounds (5,420 kg). There are fewer than 50,000 of them in the wild today.

THE ORANGUTAN

Orangutans share the rain forests of Sumatra, Indonesia, with tigers. This baby is high up in the treetops. The name *orangutan* means "person of the forest." These striking animals are covered in shaggy orange fur. Orangutans are also endangered and in need of human help.

THE AMUR LEOPARD

The Amur leopard is another big cat that lives near the Siberian tiger. Like all leopards, the Amur leopard is a superfast runner. It can reach speeds of up to 37 miles per hour (59.5 kph). It is also an amazing jumper. Some animals have been seen to jump distances of 10 feet (3 m) high and 19 feet (5.7 m) long! Amur leopards are **critically endangered**.

THE MALAYAN TAPIR

Malayan tapirs are unusual-looking animals that pad about in tiger territory. They have long and fleshy noses. Tapirs have lived on Earth for at least 50 million years. Tapirs are threatened by habitat loss and hunting, just like the tiger.

UNDER THREAT FROM HUNTING

Poaching is the illegal killing or capture of animals. Poachers hunt and kill tigers because they can sell their skins, bones, and other body parts for a lot of money. Poaching is one of the biggest threats that tigers face.

AGAINST THE LAW

The illegal wildlife **trade** is the illegal trade of wild plants and animals. This multibillion-dollar business is said to be worth between $50 and $150 billion dollars a year. The tiger is one of its many victims. Almost every part of a tiger has been found in illegal wildlife markets, from its whiskers and tails to its teeth and claws.

Kids on a Mission

@luke_animal_sos1

I held a fundraising party for tigers this year. We donated all the money we made to a wildlife charity that protects tigers.

TIGER POWER

Tiger body parts have huge value to many people. They believe the animals have powerful healing powers. Everything from the tiger's fur to its bones are ground up and made into traditional medicines. These are then sold across the world. Medicines made from tigers are very popular in China and Southeast Asia. There, they are sold in markets and stores. Medicines made from tigers are also sold in Europe, North America, and other places around the world.

"Our tiger reserves are like open treasures. For people into illegal trade, the tiger is a valuable thing for them."

Dr. Raghu Chundawat, India's leading tiger conservationist

WORTH A LOT

Tiger skins also have great value. They are sometimes made into rugs. Other times, they are hung on hunters' walls as **trophies**. Tiger skins are also bought by people in Asia, who view them as **status symbols**. In some Asian countries, tiger meat is also eaten as a special food.

▲ These Bengal tigers live in Pench National Park, in India. Tigers should be safe in reserves like these. Sadly, poachers kill them even here.

SPOTLIGHT ON POACHING

Throughout history, tigers have been killed by hunters. They were first hunted because people were afraid of them. Tigers hunted humans in **prehistoric** times, because they were a source of food. Humans in turn hunted and killed the tigers to protect themselves and their families. Later, hunting tigers became a sport. People wanted to capture tigers to display them to others. They also killed them as a sign of their strength. Today, people kill tigers for money.

▼ Tiger skins are so valuable that poachers go to any length to get them—including hunting tigers in protected reserves.

KILLING TIGERS

Poachers use different ways to kill tigers. They use guns and poison. They even hack them to death with knives. It is a horrible death for the tiger. Poachers often make camp in the middle of forests in which tigers live. There, they set up traps for the tigers made from ropes and nets. They check the traps every day in the hope that there will be a tiger inside. Poachers even hunt tigers in wildlife reserves, where they should be safe from any threats.

PEOPLE ON A MISSION: TIGER PROTECTORS

The job of a wildlife officer is to protect tigers from threats in their habitat. That includes poaching. Wildlife officers patrol areas they are in charge of every day to check on the tigers that live there. They usually travel between 6.2 and 9.3 miles (10 and 15 km) a day.

During a patrol, officers search for information about tigers. They check for signs of tigers, including paw prints or droppings, called scat. Officers use the Spatial Monitoring And Reporting Tool (SMART) to record findings. This is software that is used to keep track of what is happening to tigers. If officers have a cell phone, they use a SMART app to record their findings as they patrol an area.

The SMART information is stored in a central system, so it can be shared with other officers.

While on patrol, officers will check for signs of illegal activity, including poaching and illegal logging. They log any signs of illegal activity in their SMART system. Officers remove any snares and traps that they find during their patrols. They also take action against poachers and illegal loggers. Unfortunately, because there are many more poachers than wildlife officers, the poachers are difficult to catch. Officers are also in danger from poachers. Officers who try and catch poachers are often attacked. As a result, many have lost their lives trying to protect tigers.

Tiger skins and medicines made ▶ from tigers may be sold in markets.

BABIES NEED THEIR MOMS

Tiger poaching is a huge problem. Tigers are now so endangered that the death of just one tiger can have a terrible effect. If a female tiger with cubs is killed, her cubs will almost certainly die without their mother to protect them. Tiger cubs depend on their mothers for their survival.

MOM TO BE

A tiger mother is pregnant for around 100 days before she gives birth to her cubs. Before the babies are born, she finds a safe place to give birth. On average, tiger mothers have between two and four cubs.

MOMS AND BABIES

The bond between a cub and its mother is incredibly strong. A tiger mother will do everything she can to keep her babies alive. When the cubs are newborn, the mother **nurses** them day and night.

KEEPING CLEAN

A tiger mother keeps her cubs clean by licking them with her tongue. She snuggles them to comfort them, but she will also scold them if they do something wrong!

KEEPING SAFE

A mother tiger leaves her cubs only to hunt. While their mother is gone, the cubs stay in a sheltered place. There, they are more likely to be safe from danger. Tiger cubs are vulnerable to predators, including other big cats and people.

PLAYTIME

Just like human children, tiger cubs learn through play. They play with their brothers and sisters to build strength. Playing also teaches them to pounce and bite. When cubs are a few months old, they leave the safety of their shelter to follow their mom as she hunts for prey.

CHAPTER FOUR

CLIMATE CHANGE THREATS

It's not just humans who are threatened by climate change. So, too, are tigers. As Earth heats up because of global warming, the planet is experiencing more extreme weather. And it is affecting tigers.

If extreme weather causes tiger ▶ grasslands to become incredibly dry, tigers are also at risk of wildfires.

WHAT IS CLIMATE CHANGE?

Climate change is a change in temperature or weather in an area over a long period of time. Climate change has always happened slowly throughout Earth's history, due to natural factors such as changes in the sun and **geological** activity. However, the climate change we are experiencing today is largely caused by human activities. These include burning fossil fuels, which releases gases into Earth's **atmosphere**. These gases circle the planet like a blanket, causing it to heat up. This is called global warming.

*"I want to contribute toward **preserving** our natural habitats and save tigers, so that my next generation can live in harmony with nature."*

Debmalya Roy Chowdhury, WWF

FLOODS ON THE INCREASE

Most tigers live in parts of the world where there is a rainy season and a dry season. Climate change is affecting these seasons. It is causing huge amounts of rain to sometimes fall, causing terrible **flooding**. Floodwaters can rip away plants and trees, damaging natural habitats.

Gases pumped into ▶ Earth's atmosphere are threatening tiger habitats by causing global warming, resulting in climate change.

DROUGHTS AND RISING SEAS

Climate change is also causing delays in the rainy season. Sometimes, little or no rain falls, too. If the rains are very late one year or do not come at all, the land becomes **parched**. A long period with little or no rain is called a drought. Droughts are a serious problem for tigers because they need to be near water. Climate change is also causing **sea levels** to rise, which is threatening some tiger territory in coastal areas.

SPOTLIGHT ON SEA LEVELS

Rising sea levels are a major problem for tigers in the Sundarban islands. These islands, near the mouth of the Ganges River, have a lot of **mangrove** swamps. As sea levels rise, the water is covering more and more swamp land. Around 100 Bengal tigers live in the mangrove forests of the swamps. They are all threatened by rising seawater.

LOST ISLANDS, LOST TIGERS

As the sea level rises, the tigers' habitat is being swallowed up by the water. This is leaving the tigers with less and less land on which they and their prey can live. The WWF believes that if sea levels continue to rise at their current rate, by 2070 the precious mangrove forests of the Sundarban islands will have disappeared altogether.

PEOPLE ON A MISSION:
FIGHTING CLIMATE CHANGE

Many people are helping tigers by studying climate change and trying to stop it. They work in environmental careers. These jobs are just as important to the tiger as those that involve working directly with the animal.

Climate and environmental scientists conduct research on climate change. They study how it is affecting Earth. They note the effect of climate change on both the environment and the animals that live within it. For example, the scientists carefully monitor sea levels and how they are affecting the mangrove swamps in which Bengal tigers live. They then tell conservationists about their findings.

Environment lawyers work on climate change cases. They may be employed by organizations and governments that try and help protect tigers. Lawyers will be used to defend tigers and their environment. They will act against businesses or individuals that break tiger-protection laws.

Campaigners also raise awareness of climate change and the effect it is having on tigers and other wildlife. They work in organizations such as charities. They help run campaigns that alert governments, other organizations, and the public about the dangers of climate change.

TIGERS NEED WATER, AND THEY LOVE IT, TOO!

It's not just the tigers of the Sundarban islands that love water. All tigers are water-loving animals. They usually live close to a lake, pond, or river. They need water to drink. They also cool down in it.

SEE ME SWIM!

Tigers are great swimmers. They can swim across rivers and can swim for distances of up to 18 miles (30 km), even in strong currents.

ISLAND HOPPING

The tigers that live in the mangrove forests of the Sundarban islands even swim from one island to the other!

TIGERS TAKING SHORTCUTS

Tigers take smart shortcuts as they patrol their territory—by swimming across it!

KEEPING COOL

When temperatures are very hot, tigers cool off in water.

GOLD-STAR SWIMMERS!

Tigers have partly webbed toes that increase the surface area of their feet. That helps the tigers swim quickly.

CHAPTER FIVE

OTHER THREATS TO TIGERS

Although habitat loss, hunting, and climate change are the main threats to tigers, they also face other dangers. Tigers are big business. Some people are **breeding** these animals in **captivity**—on tiger farms. They do so to help feed the illegal wildlife trade with tiger products. Tigers are also being kept for people's pleasure—as pets!

FARMING TIGERS

The WWF estimates that as many as 8,000 tigers live in captivity on tiger farms in East and Southeast Asia. Organizations such as the WWF are working with governments in Asia to control and stop tiger farms.

"Through storytelling, filmmaking, and social media, we can show their (tigers') gentle side followed by their importance in our ecosystem."

Suyash Keshari, wildlife filmmaker and conservationist

FUN (BUT NOT FOR THE TIGERS)

People's fascination with tigers has a long history. In ancient Rome, gladiators used to fight tigers. People were even fed to the cats—they were eaten alive for the audience's amusement. And in the nineteenth century, magnificent tigers were put on display in circuses.

◀ Today, there are more tigers in captivity than there are in the wild.

NOT PERFECT PETS

Today, people are still drawn to majestic tigers. Some people are so fascinated by them that they even keep them as pets! Across the United States, tigers are being kept in people's homes. They may keep them in their backyards, basements, and even in city apartments. Tigers are not ordinary cats that can be petted and kept happily at home. Cute cubs grow into dangerous animals that need to live in the wild.

It's not all bad news for tigers in captivity. In fact, keeping tigers captive can help protect them. Tigers kept in zoos are protected from poachers, habitat loss, and climate change. In zoos, scientists can study tigers, too. By finding out more about these endangered animals, they can help protect them in the wild.

CUBS IN CAPTIVITY

Another benefit to keeping tigers in zoos is that they breed well in captivity. It is safer for cubs to be born in captivity than it is in the wild. They are not in danger from predators or poachers. However, the downside of breeding tiger cubs in zoos is that these young animals will never be prepared for the wild. If they were ever released into the wild, they would not survive for long.

▼ Zoos around the world can help the tiger. By breeding tigers in captivity, zoos can keep alive species that may die out in the wild.

PEOPLE ON A MISSION: TIGER SCIENTISTS

Zoologists are scientists who study tigers in the wild and also in zoos. They also conduct research in the lab, universities, and other organizations. Zoologists may help educate people who visit zoos. They explain the threats that tigers face. They also work in wildlife reserves. There, they help check tiger populations and study their behavior.

Zoologists study tigers to find out more about them, such as what they eat and how they raise their young. They study what diseases the animals suffer from. They learn how issues such as climate change are affecting tigers. They pass on the information they find to conservationists, which helps them understand what is happening to tigers.

Wildlife forensic specialists are scientists who study wildlife crime scenes. For example, if wildlife officers discover a poaching crime, they may call in the help of a wildlife forensics specialist. This specialist uses forensic science to gather **samples** from the field. They later analyze them in a lab. They also examine the crime site to find out more about crimes committed there, such as poaching. This evidence is then used in cases against people who have committed illegal acts.

OTHER BIG CATS IN DANGER

Around the world, other big cats are in danger. Many of them face similar challenges to the tiger. They are threatened by habitat loss and hunting.

THE CHEETAH

Cheetahs live in Africa. These big cats are endangered because of hunting. Humans are also pushing farther and farther into their territory. That is reducing the land that cheetahs can live on. Climate change is also destroying their habitat. Cheetahs **reproduce** slowly. That means that if they are threatened by activities such as hunting, they are more at risk of extinction.

THE LION

Lions are magnificent animals that live in Africa on its grasslands. Like the tiger, these big cats are at risk because farmers have taken over a lot of their territory. Lions are also hunted, and their body parts are sold in the illegal wildlife trade.

THE SNOW LEOPARD

Snow leopards are mysterious creatures that are found in central Asia. These cats are threatened by humans moving into their territory, who want to use it as farmland for their livestock. Climate change is also damaging the cat's habitat. Like the tiger, snow leopards are hunted by humans.

THE JAGUAR

Jaguars are mainly found across the Amazon rain forest, in South America. Like the tiger, they like to live near water. These cats are threatened because of habitat loss. Their forest home is being cut down for timber and cleared for farmland. Jaguars are also hunted by humans for the illegal wildlife trade.

WHAT'S NEXT FOR THE TIGER?

The future for tigers is uncertain. Countries that have set up tiger reserves face huge challenges. Tiger farms that feed the illegal wildlife trade are continuing the trade in tiger body parts. And poachers still hunt tigers, despite the best efforts of wildlife officers to stop them.

HELP FROM CONSERVATIONISTS

Tigers need more conservationists to fight for them. If more people become involved with tiger conservation, they can help raise awareness of threats to tigers. They can educate and work with local people to reduce poaching and farming. They can monitor tiger populations and fight to protect their territories.

The threat to tigers is real. But we should also be encouraged by what conservationists have already achieved. In India, tigers are making a comeback! The number of wild tigers more than doubled between 2006 and 2018. It is now estimated that between 2,600 and 3,350 tigers live in India. That's around three-quarters of the world's population!

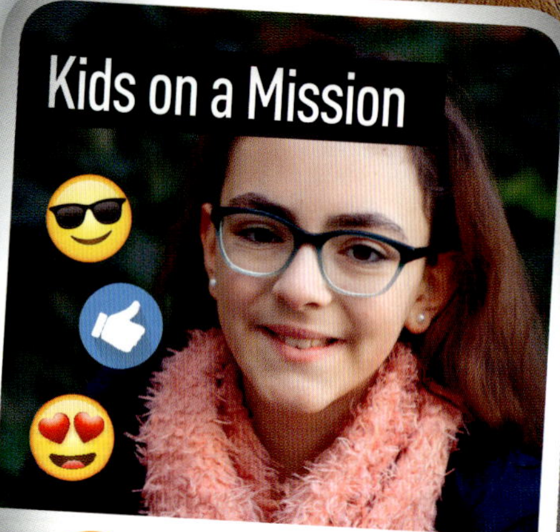

I'm going to train to become a zoologist when I grow up. Now that I know what is happening to tigers, I want to do everything I can to help.

"India now has 50 different tiger reserves. The next generation will take advantage of these success stories and build new success stories."

Dr. Raghu Chundawat

HELP FROM ORGANIZATIONS

The WWF is one organization fighting to increase tiger numbers. It is campaigning for more tiger habitat to be protected. It wants trees replanted in forests that have been cut down. It is fighting to create tiger corridors. These are protected strips of land between tiger reserves and their natural habitat. Tigers can use the corridors to safely travel from one area to another.

HELP FROM ALL OF US

There is a lot, too, that ordinary people everywhere can do to help the tiger. If we all try to live **sustainably**, we can help reduce the threat of climate change. If we all raise awareness of the illegal wildlife trade, we can help stop it. If we all take part in events such as World Tiger Day, more and more people will become aware of the threats to tigers. If we all join together, we can save the tiger. That will mean that this ancient and incredible animal will still be a big cat king for many years to come.

MAKE IT YOUR MISSION: A CAREER IN CONSERVATION

Why not make it your mission to build a future career in conservation? That way, you could make a huge difference to tigers and other endangered animals. On pages 44–45, you'll find some of the conservation careers that are possible. And here are things you can do right now to prepare for a career in conservation and help save tigers.

GET SET TO STUDY

Working hard at school now will pay off if you want a conservation career. Pay attention in your science, English, and geography classes. Languages are also useful. If you have a job oversees, being able to speak a foreign language is helpful.

GET CONNECTED

Connecting with other people who care about wildlife is a great way to feel empowered about saving endangered animals. You will find some of the organizations you can join on page 47. By connecting with other people who are concerned about the environment, you'll find out what you can do to help. Being involved with conservation organizations and charities now will also help you when you apply for jobs in the future. It will show that you have been interested in conservation from an early age.

TAKE ACTION!

You have learned about some of the actions that help tigers from the Kids on a Mission features in this book. Here are some more ideas for other activities that can help protect tigers:

- Adopt a tiger! Ask your parents to help you check out the WWF site for information about adopting tigers.
- Choose FSC-certified or recycled tissue and toilet paper. These products are not made from paper from trees cut down in rain forests.
- Ask your parents not to buy products that contain palm oil. A lot of tiger territory is cut down to make room for palm oil farming.
- Ask your parents to buy coffee that is Rainforest Alliance certified. That means that no rain forests were destroyed in its production.
- Ask your school to hold a fundraising event to raise money for an organization such as WWF, so they can help protect tigers.
- Start a blog about tigers and the dangers they face. Educate as many people as you can.

▲ If you join organizations such as WWF, you'll find out what other people are doing to help protect endangered species, too.

GET EXPERIENCE

Any form of work experience in wildlife care will be useful for a future conservation career. Why not apply to local zoos, charities, and conservation organizations as soon as you are old enough?

Working in a zoo can be a great ▶ way to get experience that will help build a future career in conservation.

CAREERS IN CONSERVATION

Today, conservation is a huge growth area, as more and more people turn their focus to concerns about the environment. There are a large number of conservation jobs available. Here is information about just some of them. You can find out about other conservation jobs on page 47 of this book.

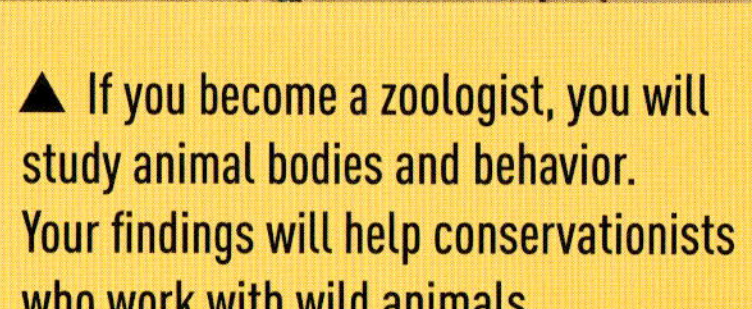

▲ If you become a zoologist, you will study animal bodies and behavior. Your findings will help conservationists who work with wild animals.

WILDLIFE MANAGER

If you love the idea of working with animals, you could become a wildlife manager. In this role, you would check on the health and well-being of animals in a natural area. You would also take care of the animals' habitat. Another important job of a wildlife manager is to educate the public about animals and the need to protect them. A degree in environmental studies is usually needed.

WILDLIFE TECHNICIAN

Wildlife technicians study wildlife in its natural habitat. In this role, you would closely watch animal numbers and animal behavior. You could work for resource companies, environmental firms, and governments. You'd help make sure that wildlife protection laws are followed. A degree in wildlife management is usually required.

ZOOLOGIST

Becoming a zoologist is a great step to take to help protect animals. In this job, you would study animal biology and how animals behave. You could work for zoos, universities, governments, and research organizations. Many zoologists travel for their work to study animals in their natural habitats. A biology degree is needed.

PUBLIC EDUCATION AND OUTREACH SPECIALIST

In this role, you would teach people about wildlife and its habitats. Giving talks at schools to educate children would be part of your job—you'd tell them about the importance of protecting wildlife. You might also give talks to people in organizations. That could include companies interested in helping wildlife. A degree in environmental studies is needed.

PUBLIC RELATIONS OFFICER

If you like raising awareness about important issues, this could be the role for you. Public relations officers manage communications between organizations and the public. In this job, you could work for charities, wildlife organizations, and governments. You would help run campaigns to make people aware of wildlife issues.

JOBS IN ENVIRONMENTAL SCIENCE

There are many different roles in environmental science. For example, you could be a climatologist or a conservation scientist. A degree in environmental science is usually required for these jobs. If you want to become an environmental lawyer, you will need a law degree with a specialization in environmental law.

◀ By working in public relations, you'll help raise awareness of the issues that tigers face.

GLOSSARY

adaptation changing to deal with new conditions

ambush a surprise attack

atmosphere the blanket of gases that surround Earth

biologist a scientist who studies living things, such as plants and animals

breeding mating to produce offspring

captivity kept in an enclosed space, not in the wild

communications the sharing of messages or information

conflict a struggle or difficult encounter

conservation protection of the planet

conservationists people who try and protect the planet

critically endangered facing an extremely high risk of extinction in the wild

enclosures spaces with walls or fencing in which animals are kept

endangered facing a very high risk of extinction in the wild

environments natural places where plants and animals live

extinction dying out

flooding the covering of normally dry land with a large amount of water

food chain the animals and plants that are connected because they eat or are eaten by one another

forensic related to the scientific investigation of a crime

fossil fuels fuels formed from the remains of plants and animals that lived long ago

geological related to Earth's structure

grasslands large open areas covered with grass and few trees

habitats places in which plants and animals live

mammals warm-blooded animals that feed their young with milk from their bodies

mangrove a tree or shrub that grows in tropical coastal areas

minerals substances found in rocks, such as gold and silver

mines openings in the ground from which resources such as minerals are taken

nurses feeds with milk

organisms living things such as plants and animals

parched dried out from heat and lack of water

plantations areas on which crops such as coffee, sugar, and tea are grown

populations all the people and other animals living in certain areas

predators animals that hunt and eat other animals

prehistoric before recorded history

preserving keeping alive or safe

prey animals that are hunted and eaten by other animals

rain forests thick forests of tall trees found in tropical areas that get a lot of rain

reflects bounces back off

reproduce to create babies or young

resources things that people use

samples small amounts for testing

sea levels the average levels of the surface of the ocean

sense related to the five senses of sight, hearing, smell, taste, and touch

species a type of plant or animal

stalk to hunt quietly

status symbols things that people believe make them seem rich and powerful

sustainably doing something in a way that does not harm the environment

territory an area of land that an animal regards as its own and that it may defend from other animals

timber wood from trees that have been cut down

trade the exchange of goods for money

trophies prizes

FIND OUT MORE

BOOKS

Adams, Sabrina. *Endangered Species and Our Future* (Spotlight On Our Future). Rosen Publishing, 2022.

Amstutz, Lisa J. *Tigers on the Hunt* (Searchlight Books). Lerner Books, 2018.

Modany, Angela. *Animal Encyclopedia: 2,500 Animals with Photos, Maps, and More!* (National Geographic Kids). National Geographic Kids, 2021.

WEBSITES

Discover more about careers that help fight climate change at:
www.bestcolleges.com/blog/climate-change-jobs

Find out what Big Cat Rescue are doing to help raise awareness of threats to tigers and end the illegal wildlife trade:
https://bigcatrescue.org

Discover the ultimate guide to careers in conservation at:
www.conservation-careers.com/15-key-conservation-jobs-ultimate-guide-for-conservation-job-seekers

Hear directly from people working in conservation. Find out what they have to say about a career in conservation at:
www.conservation-careers.com/conservation-jobs-careers-advice/how-to-get-a-job-in-conservation

Discover careers in environmental science at:
https://jobs.environmentalscience.org

Find lots of amazing wildlife careers and what they involve at this useful site:
www.thebalancecareers.com/careers-with-wildlife-125918

Discover what the world's leading wildlife organization, the WWF, is doing to help tigers and how you can get involved:
www.worldwildlife.org/species/tiger

Publisher's note to educators and parents:
All the websites featured above have been carefully reviewed to ensure that they are suitable for students. However, many websites change often, and we cannot guarantee that a site's future contents will continue to meet our high standards of educational value. Please be advised that students should be closely monitored whenever they access the Internet.

INDEX

ABOUT THE AUTHOR

Sarah Eason has written a number of books on animals and the dangers they face. In researching this book about tigers, she has learned how magnificent these animals truly are and why everyone should try and help save them.